Teaching Digital Media in an Open Source World

by Mark Page-Botelho

ISBN 978-0-6151-8716-7

Published by Mark Page-Botelho

Email: mpage@netpb.com

Table of Contents

Introduction

Inspiration

During the first couple of years working as a third grade classroom teacher, I was approached by the administration and asked if I would like to teach a Digital Media class. I thought about it and hesitantly accepted the offer. Even though I had never taught a Digital Media class, I believed I roughly knew what needed to be taught. This book outlines the course I designed and will hopefully give others who find themselves in a similar position a head start in designing their own digital media course.

Course Design

The course was designed as an elective, two to three hours per week, semester course to teach students how to use digital media in their everyday lives to help them present a message, while using only free open source programs. A strong social constructionist component is part of the curriculum as well as self help and discovery, using research techniques, both of which work hand-in-hand. Having students help each other is an important skill that needs to be learned and practiced. The foundation for the course was inspired by one of the only books I could find on this subject titled <u>Digital Media in the Classroom</u>, published by CMP Books, written by *Gigi Carlson*.

Relevance

Teachers are often asked by students about the relevance of whatever subject or skill being taught. I am a strong believer that giving a real-world problem with a producible product is the best motivator for students, and answers the legitimate questions poised by students regarding its relevance. The projects in this book give examples of projects that our school needed solutions for, which fulfilled the relevance question and the need to teach a skill.

Open Source

Open source programs are programs that are typically free and the programming code is available to the public. If one happens to be a programmer, you can modify the code and change an open source program to suit your needs. Using open source programs in schools allows for all students to download and use them at home, so they can maximize their learning without spending any money on proprietary programs. In addition, many open source programs have dedicated followings of programmers and are quickly updated to utilize current technology and trends.

Social Constructionism

I believe that while teaching, teachers should try to create a better world by instilling in students a desire to make the world a better place. Students model their lives after people they spend time with, which for the better part of their lives, are teachers. Also, giving students projects which have a socially positive spin on them and allows them to contribute to the world they live in, will once again instill an internal sense of accomplishment.

Online Collaboration

Online collaboration is the way of the future and as teachers we need to start requiring students to collaborate online. The course utilizes online collaboration software. This is not necessary, but beneficial to the students, and makes life much easier for the instructor.

Course Overview/Description

Overview

The Digital Media program focuses on developing technical skill and creative artistry using digital photography and imaging. The object is not to teach students how to learn proprietary software which will be out of date by the time the graduate, but rather to teach them concepts that are carried forward as software and technology moves forward.

The digital media course is instructor-led, project-based, and encourages student contributions.

Course Objectives

By successfully completing this program, students will understand how to:

- Learn the technical fundamentals of computer-based art and design.
- Be proficient in the use of digital photo manipulation software.
- Create meaningful computer based presentations

- Understand the basic concepts of digital photography and video production.
- Learn how to use online collaboration as a tool

Course Outline

- **Research Fundamentals**: Students learn techniques to help them research efficiently on the Internet.
- **Digital Image Manipulation Basics**: Get a thorough grounding using a digital manipulation program. Hands-on projects show how to select, enhance, distort, color, scale, and manipulate scanned images and artworks-or create them from scratch.
- **Presentation Software**: Build a working knowledge of the many different presentation programs.
- **Digital Video**: Create a digital video project.

Academic Application

- **Research**, students will learn techniques for searching efficiently using Internet search engines.
- **Critical Thinking**, students will problem solve using different strategies.
- **Writing**, Students will effectively communicate a message using

technology.

- *Collaboration*, students will learn to collaborate with fellow students and teachers using programs designed for online collaboration.

Prerequisites/Audience

For students, there are no specific prerequisite skills for this course other than a training to write to the *6+1 Writing Traits*, if one decides to follow the writing aspect of the course. However, the writing aspect of the course can be substituted with any writing curriculum as seen fit. Each course builds on the skills and knowledge gained from the previous project. The intended audience should be any student interested in using technology to communicated a message.

Teachers should be familiar with how to use a computer. However, the course relies heavily on students using self-discovery and collaboration with each other to learn the ins-and-outs of each program and task. This should allow a teacher with little background knowledge of the programs and hardware to teach the course.

Incorporating Everyday Technology

Having visited many schools around the world, I am still shocked at how little technology is being used in schools. When students are not in school, it seems as if in their everyday routines they use more technological items than in school, such as gaming consoles, personal media players, computers, and cell phones. By incorporating some of these items such as cell phones and personal media players teachers can easily entice students into doing projects that would otherwise be considered boring.

Platform Independence

The technology and software used for this course were platform independent. This means you don't have to insure that students at school and at home all have the same hardware. This will insure that everyone is included and time can be maximized at home.

Grading Standards

The Standards used for the course were taken from the *Arizona Department of Education*. I did a lot of research and found that they had the best well rounded standards regarding technology and it's use. The curriculum in the school in which the course was taught used essential questions to guide the curriculum. The essential question regarding technology was "*How can we live, learn and work successfully and responsibly in an increasingly complex, technology-driven society?*" Below are the standards used for the course,

STANDARD 1: *Fundamental Operations and Concepts*, Students understand the operations and function of technology systems and are proficient in the use of technology.

STANDARD 2: *Social, Ethical and Human Issues,* Students understand the social, ethical and human issues related to using technology in their daily lives and demonstrate responsible use of technology systems, information and software.

STANDARD 3: *Technology Productivity Tools,* Students use technology tools to enhance learning, to increase productivity and creativity and to construct technology-enhanced models, prepare publications and produce other creative works.

STANDARD 4: *Technology Communications Tools,* Building on productivity tools, students will collaborate, publish, and interact with peers, experts and other audiences using telecommunications and media.

STANDARD 5: *Technology Research Tools,* Students utilize technology-based research tools to locate and collect information pertinent to the task, as well as evaluate and analyze information from a variety of sources.

STANDARD 6: *Technology as a Tool for Problem Solving and Decision Making,* Students use technology to make and support decisions in the process of solving real world problems.

Open Source Programs

Implementing New Software

Changing a new and different program to replace a pre-existing program that everyone already uses is not an easy task. Every time I implement a different open source program to replace a proprietary program, I always get a lot of complaints. If you are changing programs be prepared for just about everyone to complain how the new program does not do what they want. Realize however, that what they are really saying is that they are not familiar with the program and can't get it to do what they want.

I always try to get a good handle on using any new program before releasing it in a school environment. This will give you a heads up on any potential questions your staff and students may have. It can take many months for the complaints to subside, just be patient and through self-discovery and a little help from others, everyone will become familiar with the new software and refuse to change back.

List of Open Source Program

Below is a list of software used for this course. Software over time changes and new programs appear often throughout the course of a school year. Keeping up with what is available is easy and takes only a few minutes of ones day when the need for a program arises. Remember to check for updates which can add a new functionality to a program.

Open Office, this is one of the best free office suites available today. It has essentially all the functionality of high end programs such as Microsoft's Office.

The GIMP, is a very powerful and feature rich digital image manipulation software. This is the core program used for the course.

GIMP Shop, modification add on for The GIMP program, see above. Makes the user interface identical to that of Adobe Photoshop.

PrimoPDF, Adobe standard PDF creation program.

KompoZer, this program is in the same family as the old Netscape Composer program. it is a very capable "What You See is What You Get" web page creation program.

Scribus, is an Adobe Illustrator type program. Very useful and very easy to use to help setup pages for publication.

FreeMind, is a mind mapping or brainstorming program. It is very easy to use and was used to help write this book!

Since web page address change often no address were given. If you are interested in finding these programs just type their title into any search engine to get the latest location of the software.

Implementation of Google Documents in the Classroom

Google is providing non-profit companies an opportunity to migrate their email service over to their Gmail service for free. The service includes personal unlimited email accounts, and access to their Google Docs applications, which includes a word processor, spread sheet, and presentation software. I've written down a short summary about my experience with Google Docs this semester.

Benefits

I teach a High School level Digital Media class to ESL students. My goal was to make it possible for me as a teacher and my students to have access to our school documents where ever and when ever we wanted. Some of the benefits included the price, it's free, making it available to everyone. Also, the collaboration within Google docs is amazing to say the least.

Having access to student work anywhere and at any time has been great. I no longer have to lug around stacks of papers and worry about losing them. I can work on one draft instead of having students re-print their work after revising. Both teachers and students have access to all previously done assignments. This allows the teacher as well as the students to monitor their progress. The revision feature allows one to check all revisions and whom they where done by. Track changes make it easy for students to accept or reject changes and suggestions made by the teacher. Another benefit that I found later after using the service, is that I can effortlessly check for plagiarism and cross-reference to make sure students don't share work with each other.

Another benefit to using Google Docs is that it is free. I strongly believe that schools should only use open source programs and if not available then they should use free software. This is important as it allows all students to use the software at home, giving the students the opportunity to continue working on projects at home with no added financial burden on the parents.

Google Docs allows for unprecedented collaboration with students. Having the ability to work on the same document at the same time is incredible, and it saves a great deal of time. Not only can you work on documents at the same time, but you can chat with your students as well. One feature that I haven't tried out is using the voice feature. All the collaboration features combined would make a great low cost addition to a school interested in creating a remote campus.

Some other noteworthy benefits include not having to waste money on printing supplies and no more excuses that the dog ate their paper. The spell check feature is incredible and it can auto detect which language you are typing in.

Considerations

There are a few considerations that need to be made when thinking about implementing Google Docs in the classroom. It is dependent on your Internet connection. Therefore, if you lose power or the Internet goes down at your school, you'll have no access to your documents. This can be remedied easily by having a backup plan. Also, if you have a slow connection, waiting for documents to load, especially presentation documents loaded with images, can be tedious.

I can't image what other benefits await once other Google Apps are incorporated into Google Docs. Google is constantly updating and adding programs to their already long list of free applications. My experience has been that it is well worth dealing with the small problems that occasionally pop up, and it has made my job much easier.

Projects

All the assignments listed are broken down into sections which should be completed in order. The methodology is important in order for the projects to build the skills and knowledge necessary to complete the projects. It is very important to realize that no one teacher will have the background knowledge necessary to teach the programs used, therefore it is important for the teacher to let the students help each other.

Research Oriented

Researching how to use programs is important, and helps the students become dynamic learners. There are many different ways to accomplish the tasks listed. Many of your students will not like the fact that they have to figure out how to fulfill the tasks themselves. Be strong and don't show them how to do it, but rather help them find the answers! All the programs uses are well documented on the Internet. Using search engines, and others in class is a great way for students to learn. Allow students to wander and talk with others about how to accomplish tasks. Do not consider this cheating, as long as the content is unique. Learning the skill is the important factor.

Fair Use of Images

One aspect that is often overlooked is the fair use of images. The rule of thumb is an image has to be modified beyond recognition in order to be considered a new image. I found it best to direct students to royalty free image sites, or to capture their own images using digital cameras or their cell phones.

Saving Images

Students will typically work on an image over many days. When they are working on the images, have them save their work in whatever the default image type is for the program. For example, The GIMP using a file type of XCF, Adobe Photoshop uses PSD. Saving in the default file type will allow students to retain layers so the images can still be easily manipulated once re-opened.

Once an image is completed, have students save the image as a JPG or GIF. JPG images are great for real life photography, while GIF is great for artwork. GIF images are proprietary and recently there has been talk about the fact that there might be a charge related to their use. This seems unlikely, but one should keep it in mind. PNG is an alternative which is widely used and works with all web browsers.

Shared Folder Storage

It is important for your class to have access to a shared folder, where you, the teacher, can access student work. Make sure that individual folders are created and/or that an agreed upon naming convention for files is strictly enforced. Also, many of the projects utilize Google Docs for sharing reflections and written work. This is not necessary, but makes life a lot easier for everyone. In order to utilize Google Docs, make sure that all students create a free Gmail email account. I had all students start the term by sending me a mandatory email.

I have listed the extra credit projects first. Make sure your students understand them as they will make everybody's experience easier.

Extra Credit Projects

Below are some extra credit assignments that I used to help bring out some collaboration between students who needed a little external motivation to help their classmates.

Extra Credit Project: Pay Back!

Warning! Only the helper gets extra credit, not the person who was helped!

If someone asks you for help, or you see someone in need of help, help them. If they agree they can email your teacher with the following information;

- Date
- Name of Helper
- Name of person who received help
- What was the problem? (Complete Sentence)
- How did the person help you? (Complete Sentence)

Extra Credit Project: Coffee Break!

Find one skill dealing with a project that we have worked on, and share it with the class. You can have use of the projector or any other tool you need to help you demonstrate your helpful idea.

When done create a Google Doc on what you covered in your Coffee Break and share it with your teacher.

List of Projects

The following are the projects preceded by a short objective of the project.

Project: Internet Research Game

Description: Students learn about different types of tools available on the Internet for research. This activity is key to all other assignments in this book.

Materials Needed: Internet Connectivity.

Stated Objective: This project involves learning or refining your Internet research skills.

Step One:

Prior to starting the game choose three or four random different items. Let the students know that they will have a race on the Internet searching for different items using any means they like, but have them somehow keep track of the web page they used and the work or phrase they used. They can play as individuals or teams, you decide. Next, have students search for the different items one at a time. You can give prizes, which always ups the ante for participation.

Once finished playing, discuss with students which search engines they used. Have them explain why they used that particular engine. Then discuss the difference between a search engine and a directory. Finally, discuss what terms or phrases they used. Talk about the use of quotes and operators such as the "+" sign.

Step Two:

Have students create a two column report with each web page or search engine they used on the left side, and corresponding search phrase or word they used on the right side.

Reflection:

1. What did you learn about searching on the Internet in this activity?

2. Are all search engines the same? Give an example of a similarity or difference.

Project: Who Am I?

Description: Students use skills learned to create images which represent them, and put them together to create a presentation.

Materials Needed: Presentation software.

Stated Objective: This project involves creating a digital presentation about your personal culture.

Step One:

Have fun looking for items that represent the student, such as favorite sport, food, movie, actor, and book. Have students copy the found image and URL of the web page it was found on.

Step Two:

"What defines you?" Have students write a short description for each attribute that represents who they are. Each student can have a different number of attributes such as political leanings, religion, outlook on life, or tastes in music or food.

Step Three:

Have students create a GIMP image file for each of their written descriptions using one of the images found on the Internet. Each slide must contain a background and items that represent their culture. Have students use each of the listed techniques on a different image.

Base for Images. Have students use GIMP to create a graphic file with following attributes:

- Width 500 pixels
- Height 500 pixels
- Resolution 72 dpi
- Mode RGB

Copy & Paste Objects. Have students find a new background image. Then create new image in GIMP and paste new background onto it. Next paste one of the found objects into the new image as a new layer. Then anchor all the objects together.

Enlarge Object. Enlarge one aspect of an object. For example mask and copy part of a house such as a window or door, then paste and enlarge it.

Color Channels. Transform the color of the object to portray your feeling about it (i.e. blue=sad, yellow=scared, red=angry) using a color filter.

Step Four:

Have students create a presentation using Open Office or MS Word. Have them combine their written descriptions and corresponding image on the slides. For a bonus add music to the presentation. Make sure that they use some type of slide animation. Have them save the presentation in the share folder with (their name + culture).

Reflection:

1. Create a Document with three rows and three columns that explain the Highs, Lows, and what you learned while using the Software (GIMP), Content (presentation), and the Process (how the project evolved). 2. Email to reflection to the teacher for grading.

Project: School Logo

Description: Students research logos and create one for their school. This is an application project where students use the skills they have learned thus far.

Materials Needed: The GIMP program.

Stated Objective: In this project you will create a new professional looking school logo.

Step One:

Have students search the Internet for professional school logos. Have them share their findings with their neighbors. A good place to find examples is TravelingEducator.com. One can also do a search on Google for "international schools".

Step Two:

Have students brainstorm and find items (graphics) that have to do with their school. Have them think about characteristics, emotions, and any other details about the school.

Step Three:

Find images for each of your details that they found in the previous activity and modify appropriately.

Step Four:

Using GIMP have students combine their images to create a "professional looking" school logo. The image should be a JPG image, 300 pixels in width x 300 pixels in height, or 750 pixels in width x 136 pixels in height for a more square Graphic. Have students add Text Information to the images, such as a school slogan, or school name.

Project: Wacky Ad

Description: This project was designed to give students a fun application project.

Materials Needed: The GIMP program.

Stated Objective: In this project students will practice the image manipulation skills they have learned so far, and will create a funny or weird image using contradicting images.

Step One:

Have students search the Internet for creative advertisements or commercials, then share their findings with their neighbors.

Step Two:

Have students find an item that they have an interest in. Have them think about characteristics, emotions, and any other detail about their object.

Have students find the opposite, contradicting ideas or thoughts, emotions, and any other detail about their object.

Have students create a table to depict their findings and email it to their teacher.

Step Three:

Have students find images for each of their details that they found in the previous activity.

Step Four:

Using GIMP have students combine their images to create a Wacky Ad! Their ad must contain a slogan.

Project: Gump

Description: Once again a fun application project.

Materials Needed: The GIMP, Internet Access.

Stated Objective: For this project you will create fictional photos staring you posing as a character in a historical photo.

Step One:

Have students find three advertisements where the subject demonstrates expression. Have students share their findings with the class or neighbor.

Step Two:

Have students create or find an expressive digital photo of themselves.

Step Three:

Have students find different images of famous events (i.e. Pope visit to South American) and insert themselves into them by replacing their face with someone in the photo. They should play off of the expression of your stock photo.

Step Four:

Create a slide show of their new expressive photos. Have students save their work as GUMP-(their name).

Project: Brochure

Description: Students create a paper brochure that a school can use to recruit new teachers with. This project can easily be modified for another purpose.

Materials Needed: Open Office, tri-fold brochure template available online at

http://documentation.openoffice.org/Samples_Templates/User/template/index.html

Stated Objective: For this project you will create a paper tri-fold brochure on a topic chosen by your teacher.

In this assignment students will be helping their school recruit new teachers. Unfortunately, there is not much information about their school so many teachers pass up the opportunity to work and live at their school and hometown. In order to help recruit new teachers, they will be creating brochures about living and working in your town and at your school. Their goal is to make a tri-fold brochure using graphic images and text to help convince perspective teachers to move to your location.

Step One:

Have students put themselves in the shoes of a new perspective teacher who just got an offer to move to your home town to work at your school! The only problem is they never heard of the place. Search the Internet and search for ONE of the items below and see how much information they can come up with.

If you can think of something that you believe a perspective teacher would like to know that is not listed, ask your teacher if you can do it.

- Medical Facilities
- Sport and Leisure
- Travel and Sightseeing
- Shopping
- Services
- Auto
- Restaurants or Cafes
- Housing & Construction
- Photos

Reflection:

We're you able to find enough information about your school to convince you to move here if you knew only what you found out about it?

What do you think is important to know that was missing? Include at least three links that you looked at.

Have students submit a reflection via Gmail to your teacher.

Step Two:

Have students collect images using the Internet and/or find real photos that they or someone else has taken. they must collect at least 5 images. Crop the images and fix them up using the techniques they learned about previously (filters such as sharpen, channel mixer, cropping, and layering).

Have students store their images in a shared folder for review.

Have students label them Brochure1, Brochure2, ect. Mandating a naming convention is an important skill to learn when dealing with many images that need to remain organized. Make sure they are saved as JPG's. This insures students become familiar with the JPG file type, which is widely used on the Internet and by other programs. The images cannot be larger than 300 pixels in any given direction. This will insure students learn how to resize images.

Step Three:

Have students write informational text using a persuasion format, minimum three paragraphs. Have them submit rough draft via Gmail to their teacher. The final copy should be graded using 6+1!

Step Four:

Have students combine their persuasive essay and their five photos to create a tri-fold brochure in Open Office (use this tri-fold brochure template found on the Internet). Have students save their final copy as Final_Brochure in a shared folder.

Project: PodCast

Description: Students need to practice speaking aloud, which has the added benefit of having them catch their own mistakes they may have missed during the writing process. When they realize their writing sounds funny or is hard to read aloud they will have another chance to correct their work. This assignment gives them an opportunity to create a personalized project to help instill a little internal motivation (great project for second language learners).

Materials Needed: Audacity program, headphones, quiet room for recording.

Stated Objective: In this assignment you will learn how to create a simple podCast.

Step One:

Have students find at least three podcasts on any subject they like (NOTE: Have students write down the name of the podcast and the URL of where they found it).

Reflection:

In your words, what is a podcast?

What characteristics make a podcast different than paper media?

What characteristics make a podcast different than digital media?

Have students share your document ONLY via Google Docs with their teacher.

Step Two:

Interesting Note: The average amount of time that a student speaks English in an ESL school is 4 minutes per day!

Have students find an essay or a book you would like to do a podCast on. Make sure it's something they want to share with the class.

Step Three:

Have students create a podCast by reading at least three paragraphs using Audacity. Helpful Hint: Use a computer in a quiet room.

Their final podCast should be no longer than 4 minutes. Have students save audio file as (their name).mp3, and save it in a shared folder where their teacher can get to it (should be graded using school wide oral grading rubric) .

Project: Google Presentation

Description: Google Docs is the wave of the future, get in on it now!

Materials Needed: Gmail account, complete Brochure project (or a previously written essay).

Stated Objective: This project entails creating presentation using one of Google's new online applications called "Presentation."

Step One:

It's time to explore a new way of share your work with others! Have students locate and read two articles, which review or describe Google's new application "Presentation".

Reflection:

How is Google Presentation different than Microsoft or Open Office Presentation software?

What might be a drawback of the new application?

Note: Have students include URI's of articles they used.

Have students share their document ONLY via Google Docs with their teacher.

Step Two:

Have students decide on a topic such as soccer, favorite restaurant, baking a cake, (anything really). Have them collect images on any subject matter they like using real photos that they or someone else has taken. They should collect at least 5 images. Have students crop your images and fix them up using the techniques they learned about earlier (filters such as sharpen, channel mixer, cropping, and layering). Have them put the images in a shared folder where the teacher has access. Label them Google1, Google2, ect. Make sure they are saved as JPG's. They shouldn't be larger than 600 pixels in any given direction.

Step Three:

Have students write a narrative or expository essay on the subject matter they chose (minimum 5 paragraphs, three sentences each). Have students share their document ONLY via Google Docs as "Google Presentation Essay" with their teacher. The final copy should be graded using 6+1!

Step Four:

Have students combine their essay and their five photos to create a Google Docs Presentation. Have students share their document ONLY via Google Docs with their teacher.

Project: Cyberbullying Video

Description: This is a project that can help alleviate the problem of cyberbullying at your school.

Materials Needed: Windows Movie Maker, digital video camera (many cell phones have them).

Stated Objective: Cyberbullying is a growing problem in schools around the world. In this project you will create a video outlining how cyberbullying is a problem and then offering a solution.

Step One:

Have students google cyberbullying. Have them look at least at three different definitions.

Reflection:

In your words, what is a cyberbullying (Include the URLs of the three websites you used)?

Have students share your document ONLY via Google Docs as "CB-Ref" with their teacher.

Step Two:

Have students think of a situation where someone could be

bullied using technology such as cell phones, web page, or online forum (make believe, but believable).

Have students share your document ONLY via Google Docs as "CB-Ess" with their teacher.

Have students create a script using their narrative. It can be done as a fake interview if they like, or as an act in a play. Make sure they include a cast description (Should be graded using 6+1).

Have students share their document ONLY via Google Docs as "CB-Scr" with their teacher.

For the filming part, students may work with friends and choose one of the skits. To make the editing part easy, film the skit in short scenes or acts.

Using a video camera (can use camera, or cell phone) and Microsoft Movie Maker, have students film skit and edit their video. Have students include a title screen, transition screens, and a conclusion screen (should be graded using General grading Rubric).

Via email, have students list who worked on their video and what role in the production of the video they did. When done, place the video in a shared folder and email their teacher the list of roles and where the video is located for review.

REFLECTION:

Make a Google Document and list the highs, lows, and improvements about the assignment.

Have students share their document ONLY via Google Docs as "CB-HL" with their teacher.

Assessment

Assessment Weights

Class grade will be based on minor reflections or contributing projects, which will account for 50% of total quarter and semester grade. They are typically graded using the General Grading Rubric. Cumulative final projects will account for the other 50% of total quarter and semester grade, and are typically graded using the 6+1 Grading Rubric.

6+1 Writing Traits

The object of the course is to have students give a message using digital media, therefore I used the 6+1 Traits of Writing, produced by the *Northwest Regional Educational Laboratory* in Portland, Oregon. This is a great way to grade the writing aspect of the projects. There are many grading scales and rubrics which I will not go into. I used a simple 5 point scale weighted depending on if the material being graded was a cumulative final project or a minor reflection or contributing project. See example below.

Oral Grading Rubric

Many schools have implemented an oral grading rubric. I used such a rubric the for podcast project. If your school does not have such a rubric, you can find one easily on the Internet.

General Grading Rubric Example

Below is the general grading rubric used for many of the projects in this book. It is very simple and made to be subjective so it can be easily modified.

General Grading Rubric				
The object of the grading rubric is to at a glance keep track of student performance The grades attributed to this rubric should only be part of a more comprehensive grading system for Secondary Students. Creativity and Initiative are key components in this rubric				
	1 or 2, Less than 40%	3, 60%	4, 80%	5, 100%
Message / Reflection	Message not present or very vague.	Certain aspects of message are vague or difficult to understand.	Message is clear and easy to understand.	Message is clear and easy to understand. Student elaborated or added personal connections.
Presentation	Presentation distracts from message.	Few if any aspects of presentation enhance message.	Most of the presentation enhances meaning of message.	Every aspect of presentation enhances meaning of message.
Skills	No evidence of learned skill set.	Some evidence of skill set.	Skill set evident.	Multiple instances of newly learned skill set.

Comprehensive Projects

All comprehensive projects Incorporate a writing component and are evaluated using a 6+1 grading rubric.

Credit

I would like to thank my daughter Kyla for taking her vacation time to help me edit my work.

Alphabetical Index

Notes:

Notes: